The approved documents

What is an approved document?

The Secretary of State has approved a series of documents that give practical guidance about how to meet the requirements of the Building Regulations 2010 for England. Approved documents give guidance on each of the technical parts of the regulations and on regulation 7 (see the back of this document).

Approved documents set out what, in ordinary circumstances, may be accepted as reasonable provision for compliance with the relevant requirements of the Building Regulations to which they refer. If you follow the guidance in an approved document, there will be a presumption of compliance with the requirements covered by the guidance. However, compliance is not guaranteed; for example, 'normal' guidance may not apply if the particular case is unusual in some way.

Note that there may be other ways to comply with the requirements – *there is no obligation to adopt any particular solution contained in an approved document*. If you prefer to meet a relevant requirement in some other way than described in an approved document, you should discuss this with the relevant building control body.

In addition to guidance, some approved documents include provisions that must be followed exactly, as required by regulations or where methods of test or calculation have been prescribed by the Secretary of State.

Each approved document relates only to the particular requirements of the Building Regulations that the document addresses. However, building work must also comply with any other applicable requirements of the Building Regulations.

How to use this approved document

This document uses the following conventions.

a. Text against a green background is an extract from the Building Regulations 2010 or the Building (Approved Inspectors etc.) Regulations 2010 (both as amended). These extracts set out the legal requirements of the regulations.

b. Key terms, printed in green, are defined in Appendix A.

c. When this approved document refers to a named standard or other document, the relevant version is listed in Appendix B (standards). However, if the issuing body has revised or updated the listed version of the standard or document, you may use the new version as guidance if it continues to address the relevant requirements of the Building Regulations.

NOTE: Standards and technical approvals may also address aspects of performance or matters that are not covered by the Building Regulations, or they may recommend higher standards than required by the Building Regulations.

Where you can get further help

If you do not understand the technical guidance or other information in this approved document or the additional detailed technical references to which it directs you, you can seek further help through a number of routes, some of which are listed below.

a. The Planning Portal website: www.planningportal.gov.uk.

b. *If you are the person undertaking the building work:* either from your local authority building control service or from an approved inspector.

c. *If you are registered with a competent person scheme:* from the scheme operator.

d. *If your query is highly technical:* from a specialist or an industry technical body for the relevant subject.

The Building Regulations

The following is a high level summary of the Building Regulations relevant to most types of building work. Where there is any doubt you should consult the full text of the regulations, available at www.legislation.gov.uk.

Building work

Regulation 3 of the Building Regulations defines 'building work'. Building work includes:

a. the erection or extension of a building
b. the provision or extension of a controlled service or fitting
c. the material alteration of a building or a controlled service or fitting.

Regulation 4 states that building work should be carried out in such a way that, when work is complete:

a. for new buildings or work on a building that complied with the applicable requirements of the Building Regulations: the building complies with the applicable requirements of the Building Regulations
b. for work on an existing building that did not comply with the applicable requirements of the Building Regulations:
   (i) the work itself must comply with the applicable requirements of the Building Regulations
   (ii) the building must be no more unsatisfactory in relation to the requirements than before the work was carried out.

Material change of use

Regulation 5 defines a 'material change of use' in which a building or part of a building that was previously used for one purpose will be used for another.

The Building Regulations set out requirements that must be met before a building can be used for a new purpose. To meet the requirements, the building may need to be upgraded in some way.

Materials and workmanship

In accordance with regulation 7, building work must be carried out in a workmanlike manner using adequate and proper materials. Guidance on materials and workmanship is given in Approved Document 7.

Energy efficiency requirements

Part 6 of the Building Regulations imposes additional specific requirements for energy efficiency.

If a building is extended or renovated, the energy efficiency of the existing building or part of it may need to be upgraded.

Notification of work

Most building work and material changes of use must be notified to a building control body unless one of the following applies.

a. It is work that will be self-certified by a registered competent person or certified by a registered third party.

b. It is work exempted from the need to notify by regulation 12(6A) of, or Schedule 4 to, the Building Regulations.

Responsibility for compliance

People who are responsible for building work (for example, the agent, designer, builder or installer) must ensure that the work complies with all applicable requirements of the Building Regulations. The building owner may also be responsible for ensuring that work complies with the Building Regulations. If building work does not comply with the Building Regulations, the building owner may be served with an enforcement notice.

Contents

	Page
The approved documents	i
The Building Regulations	iii
Approved Document P: Electrical safety – Dwellings	1
Summary	1
Interaction with other parts of the Building Regulations	1
Requirement P1: Design and installation	2
Performance	2
Section 1: Design and installation	3
General	3
New dwellings	3
New dwellings formed by a change of use	3
Additions and alterations to existing electrical installations	4
Section 2: Application of Part P	5
General	5
Scope	5
Notifiable work	6
Non-notifiable work	7
Section 3: Certification, inspection and testing	9
General	9
Self-certification by a registered competent person	9
Certification by a registered third party	9
Certification by a building control body	10
Inspection and testing of non-notifiable work	10
Appendix A: Key terms	11
Appendix B: Standards referred to	12

Approved Document P: Electrical safety – Dwellings

Summary

0.1 This approved document gives guidance on how to comply with Part P of the Building Regulations. It contains the following sections:

Section 1: Technical requirements for electrical work in dwellings
Section 2: The types of building and electrical installation within the scope of Part P, and the types of electrical work that are notifiable
Section 3: The different procedures that may be followed to show that electrical work complies with Part P
Appendix A: Key terms
Appendix B: Standards referred to.

Interaction with other parts of the Building Regulations

0.2 Other parts of the Building Regulations contain requirements that affect electrical installations. Examples include, but are not limited to, the following:

a. Part A (Structure): depth of chases in walls, and size of holes and notches in floor and roof joists

b. Part B (Fire safety): fire safety of certain electrical installations; provision of fire alarm and fire detection systems; fire resistance of service penetrations through floors, walls and ceilings

c. Part C (Site preparation and resistance to contaminants and moisture): resistance of service penetrations to rainwater and contaminants such as radon

d. Part E (Resistance to the passage of sound): soundproofing of service penetrations

e. Part F (Ventilation): dwelling ventilation rates

f. Part L (Conservation of fuel and power): energy efficient lighting

g. Part M (Access to and use of buildings): height of socket-outlets and switches.

P1 Design and installation

Requirement P1: Design and installation

This approved document deals with the following requirement from Part P of Schedule 1 to the Building Regulations 2010.

Requirements

Requirement

Design and installation

P1. Reasonable provision shall be made in the design and installation of electrical installations in order to protect persons operating, maintaining or altering the installations from fire or injury.

Limits on application

The requirements of this part apply only to electrical installations that are intended to operate at low or extra-low voltage and are:

(a) in or attached to a dwelling;

(b) in the common parts of a building serving one or more dwellings, but excluding power supplies to lifts;

(c) in a building that receives its electricity from a source located within or shared with a dwelling; or

(d) in a garden or in or on land associated with a building where the electricity is from a source located within or shared with a dwelling.

Performance

In the Secretary of State's view, the requirements of Part P will be met if low voltage and extra-low voltage electrical installations in dwellings are designed and installed so that both of the following conditions are satisfied.

a. They afford appropriate protection against mechanical and thermal damage.

b. They do not present electric shock and fire hazards to people.

Section 1: Design and installation

General

1.1 Electrical installations should be designed and installed in accordance with **BS 7671:2008** incorporating Amendment No 1:2011.

Provision of information

1.2 Sufficient information should be provided to ensure that people can operate, maintain or alter an electrical installation with reasonable safety.

The information should comprise items listed in **BS 7671** and other appropriate information including:

   a. electrical installation certificates or reports describing the installation and giving details of the work carried out

   b. permanent labels, for example on earth connections and bonds, and on items of electrical equipment such as consumer units and residual current devices (RCDs)

   c. operating instructions and logbooks

   d. for unusually large or complex installations only, detailed plans.

Functionality requirements

1.3 Part P of the Building Regulations covers the safety of electrical installation work; it does not cover system functionality. Other parts of the Building Regulations and other legislation cover the functionality of electrically powered products such as fire alarm systems, fans and pumps.

New dwellings

1.4 Wall-mounted socket-outlets, switches and consumer units in new dwellings should be easy to reach, in accordance with Part M of the Building Regulations (Access to and use of buildings).

NOTE: Approved Document M recommends that in new dwellings only, switches and socket-outlets for lighting and other equipment should be between 450mm and 1200mm from finished floor level. Approved Document M does not recommend a height for new consumer units. However, one way of complying with Part M in new dwellings is to mount consumer units so that the switches are between 1350mm and 1450mm above floor level. At this height, the consumer unit is out of reach of young children yet accessible to other people when standing or sitting.

New dwellings formed by a change of use

1.5 Where a material change of use creates a new dwelling, or changes the number of dwellings in a building, regulation 6 requires that any necessary work is carried out to ensure that the building complies with requirement P1. This means that in some cases the existing electrical installation will need to be upgraded to meet current standards.

NOTE: If existing cables are adequate, it is not necessary to replace them, even if they use old colour codes.

P1 Design and installation

Additions and alterations to existing electrical installations

1.6 Regulation 4(3) states that when building work is complete, the building should be no more unsatisfactory in terms of complying with the applicable parts of Schedule 1 to the Building Regulations than before the building work was started. Therefore, when extending or altering an electrical installation, only the new work must meet current standards. There is no obligation to upgrade the existing installation unless either of the following applies.

   a. The new work adversely affects the safety of the existing installation.

   b. The state of the existing installation is such that the new work cannot be operated safely.

1.7 Any new work should be carried out in accordance with **BS 7671**. The existing electrical installation should be checked to ensure that the following conditions are all satisfied.

   a. The rating and condition of existing equipment belonging to both the consumer and to the electricity distributor are suitable for the equipment to carry the additional loads arising from the new work.

   b. Adequate protective measures are used.

   c. The earthing and equipotential bonding arrangements are satisfactory.

Section 2: Application of Part P

General

2.1 All electrical installation work carried out in a dwelling is subject to requirement P1, and should comply with the design and installation guidance in Section 1. Section 2 sets out:

a. the types of building and electrical installation that are within the scope of Part P

b. the types of electrical work that are notifiable and must be certified as complying with the Building Regulations.

Certification procedures are set out in Section 3.

Scope

2.2 Part P applies to electrical installations:

a. in a dwelling-house or flat, and to parts of the installation that are:

   (i) outside the dwelling – for example fixed lighting and air conditioning units attached to outside walls, photovoltaic panels on roofs, and fixed lighting and pond pumps in gardens

   (ii) in outbuildings such as sheds, detached garages and domestic greenhouses.

b. in the common access areas of blocks of flats such as corridors and staircases

c. in shared amenities of blocks of flats such as laundries, kitchens and gymnasiums

d. in business premises (other than agricultural buildings) connected to the same meter as the electrical installation in a dwelling – for example shops and public houses below flats.

2.3 Part P does not apply to electrical installations:

a. in business premises in the same building as a dwelling but with separate metering

b. that supply the power for lifts in blocks of flats (but Part P does apply to lift installations in single dwellings).

NOTE: Schedule 2 to the Building Regulations identifies buildings – for example unoccupied, agricultural, temporary and small detached buildings – that are generally exempt from the requirements of the Regulations. However, conservatories, porches, domestic greenhouses, garages and sheds that share their electricity with a dwelling are not exempt from Part P (by virtue of regulation 9(3)) and must comply with its requirements.

2.4 The scope of Part P is illustrated in Diagram 1.

P1 Application of Part P

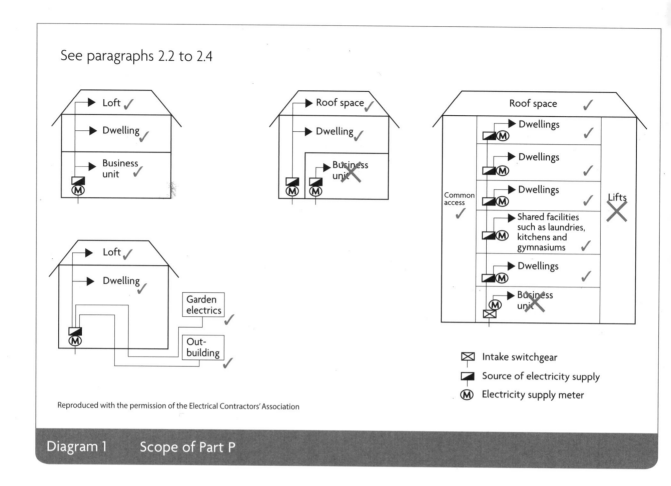

Diagram 1 Scope of Part P

Notifiable work

2.5 Electrical installation work that is notifiable is set out in regulation 12(6A).

> 12.—(6A) A person intending to carry out building work in relation to which Part P of Schedule 1 imposes a requirement is required to give a building notice or deposit full plans where the work consists of—
> (a) the installation of a new circuit;
> (b) the replacement of a consumer unit; or
> (c) any addition or alteration to existing circuits in a special location.
>
> —(9) In this regulation "special location" means—
> (a) within a room containing a bath or shower, the space surrounding a bath tap or shower head, where the space extends—
>   (i) vertically from the finished floor level to—
>   (aa) a height of 2.25 metres; or
>   (bb) the position of the shower head where it is attached to a wall or ceiling at a point higher than 2.25 metres from that level; and
>   (ii) horizontally—
>   (aa) where there is a bath tub or shower tray, from the edge of the bath tub or shower tray to a distance of 0.6 metres; or
>   (bb) where there is no bath tub or shower tray, from the centre point of the shower head where it is attached to the wall or ceiling to a distance of 1.2 metres; or
> (b) a room containing a swimming pool or sauna heater.

Application of Part P

2.6 Diagram 2 illustrates the space around a bath tub or shower tray (a special location) within which minor additions and alterations to existing circuits, as well as the installation of new circuits, are notifiable.

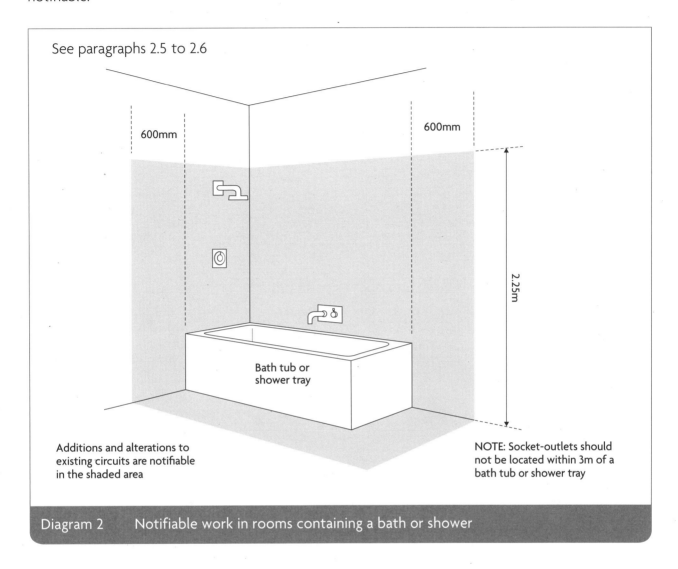

Diagram 2   Notifiable work in rooms containing a bath or shower

Non-notifiable work

2.7 Regulation 12(6A) sets out electrical installation work that is notifiable. All other electrical installation work is not notifiable – namely additions and alterations to existing installations outside special locations, and replacements, repairs and maintenance anywhere.

2.8 Installing fixed electrical equipment is within the scope of Part P, even if the final connection is by a standard 13A plug and socket, but is notifiable only if it involves work set out in regulation 12(6A). For example:

a. installing a built-in cooker is not notifiable work unless a new cooker circuit is needed

b. connecting an electric gate or garage door to an existing isolator switch is not notifiable work, but installing a new circuit from the consumer unit to the isolator is notifiable.

Application of Part P

2.9 Installing prefabricated, modular wiring (for example for kitchen lighting systems) linked by plug and socket connectors is also within the scope of Part P, but again is notifiable only if it involves work set out in regulation 12(6A).

Certification, inspection and testing

Section 3: Certification, inspection and testing

General

3.1 For notifiable electrical installation work, one of the following three procedures must be used to certify that the work complies with the requirements set out in the Building Regulations.

   a. Self-certification by a registered competent person.

   b. Third-party certification by a registered third-party certifier.

   c. Certification by a building control body.

3.2 To verify that the design and installation of electrical work is adequate, and that installations will be safe to use, maintain and alter, the electrical work should be inspected and tested in accordance with the procedures in **BS 7671**.

   NOTE: Electrical inspection and test forms should be given to the person ordering the work. Building Regulations certificates should normally be given to the occupier, but in the case of rented properties may be given to the person ordering the work and copied to the occupier.

Self-certification by a registered competent person

3.3 Electrical installers who are registered competent persons should complete a **BS 7671** electrical installation certificate for every job they undertake. The electrical installer should give the certificate to the person ordering the work.

3.4 The installer or the installer's registration body must within 30 days of the work being completed do both of the following.

   a. Give a copy of the Building Regulations compliance certificate to the occupier.

   b. Give the certificate, or a copy of the information on the certificate, to the building control body.

Certification by a registered third party

3.5 Before work begins, an installer who is not a registered competent person may appoint a registered third-party certifier to inspect and test the work as necessary.

3.6 Within 5 days of completing the work, the installer must notify the registered third-party certifier who, subject to the results of the inspection and testing being satisfactory, should then complete an electrical installation condition report and give it to the person ordering the work.

   NOTE: The electrical installation condition report should be the model **BS 7671** form or one developed specifically for Part P purposes.

3.7 The registration body of the third-party certifier must within 30 days of a satisfactory condition report being issued do both of the following.

   a. Give a copy of the Building Regulations compliance certificate to the occupier.

   b. Give the certificate, or a copy of the information on the certificate, to the building control body.

P1 Certification, inspection and testing

Certification by a building control body

3.8 If an installer is not a registered competent person and has not appointed a registered third-party certifier, then before work begins the installer must notify a building control body.

3.9 The building control body will determine the extent of inspection and testing needed for it to establish that the work is safe, based on the nature of the electrical work and the competence of the installer. The building control body may choose to carry out any necessary inspection and testing itself, or it may contract a specialist to carry out some or all of the work and furnish it with an electrical installation condition report.

3.10 An installer who is competent to carry out inspection and testing should give the appropriate **BS 7671** certificate to the building control body, who will then take the certificate and the installer's qualifications into account in deciding what further action, if any, it needs to take. Building control bodies may ask installers for evidence of their qualifications.

3.11 This can result in a lower building control charge as, when setting its charge, a local authority is required by the Building (Local Authority Charges) Regulations 2010 to take account of the amount of inspection work that it considers it will need to carry out.

3.12 Once the building control body has decided that, as far as can be ascertained, the work meets all Building Regulations requirements, it will issue to the occupier a Building Regulations completion certificate (if a local authority) or a final certificate (if an approved inspector).

Inspection and testing of non-notifiable work

3.13 Non-notifiable electrical installation work, like notifiable work, should be designed and installed, and inspected, tested and certificated in accordance with **BS 7671**.

3.14 If local authorities find that non-notifiable work is unsafe and non-compliant, they can take enforcement action.

Appendix A: Key terms

The following are key terms used in this document:

Building control body
A local authority or private sector approved inspector

Building Regulations compliance certificate
A certificate issued by an installer registered with an authorised competent person self-certification scheme, or by a certifier registered with an authorised third-party certification scheme stating that the work described in the certificate complies with regulations 4 and 7 of the Building Regulations 2010 (that is, the work complies with all applicable requirements in the Building Regulations)

Electrical installation*
Fixed electric cables or fixed electrical equipment located on the consumer's side of the electricity supply meter

Extra-low voltage*
A voltage not exceeding 50V ac or 120V ripple-free dc, whether between conductors or to earth

Low voltage*
A voltage exceeding extra-low voltage but not exceeding 1000V ac or 1500V dc between conductors, or 600V ac or 900V dc between conductors and earth

Registered competent person
A competent person registered with a Part P competent person self-certification scheme

Registered third-party certifier
A competent person registered with a Part P competent person third-party certification scheme

NOTE: *Terms defined in regulation 2 of the Building Regulations 2010

Appendix B: Standards referred to

BS 7671
Requirements for Electrical Installations
[2008 + A1:2011] (IET Wiring Regulations 17th Edition,
ISBN 978-1-84919-269-9)